5-25-85

To Blair on his 6th birthday

from

Great Grandpa & grandma B.

I want to be a fisherman

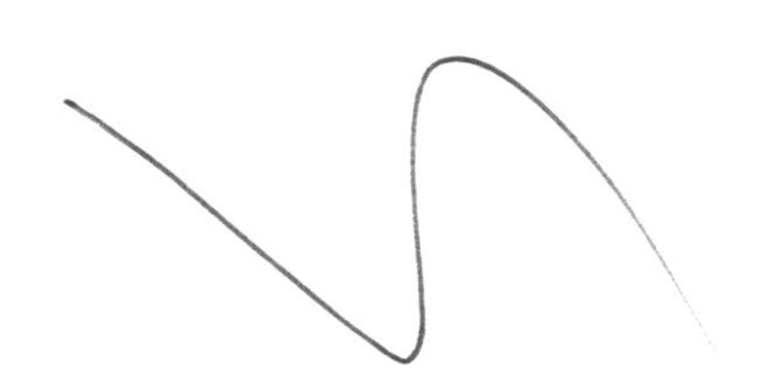

SANDRA WEINER

I want to be a fisherman

MACMILLAN PUBLISHING CO., INC.
New York
COLLIER MACMILLAN PUBLISHERS
London

AUTHOR'S NOTE

The heroine of *I Want to Be a Fisherman* is Christine Vorpahl, the daughter of my Long Island neighbors, Stuart and Mary Vorpahl. For several generations the Vorpahl family have earned their livelihood by trap fishing—an occupation handed down from the American Indians, and now a vanishing way of life.

The words and photographs that make up this book are the result of two summers I spent documenting the adventures and challenges of their exciting craft.

In doing this book, I learned not only about trap fishing and mending nets, but how each member of this family shares in the day-to-day problems of living and working together.

The photographs on the title page and pages 1, 25 and 58 are from *The Dhammapada: The Sayings of the Buddha*, translated by Thomas Byrom, photographs by Sandra Weiner.
Macmillan Publishing Co., Inc.
866 Third Avenue, New York, N.Y. 10022
Collier Macmillan Canada, Ltd.

Printed in the United States of America

10 9 8 7 6 5 4 3 2 1

LIBRARY OF CONGRESS CATALOGING IN PUBLICATION DATA

Weiner, Sandra.
I want to be a fisherman.

SUMMARY: Text and photographs introduce a trap fisherman and his daughter who practice an ancient form of fishing off the Long Island coast.
1. Fishermen—New York (State)—Long Island—Juvenile literature. 2. Fish traps—Juvenile literature. [1. Fishermen. 2. Fish traps] I. Title.
SH331.15.W44   639'.2'023   76-48088
ISBN 0-02-792520-X

To the memory of
SUSANNA and SILVANA

Chapter One

Tomorrow I will be eleven years old.
My father has promised to give me
a new knife for gutting fish.
My father is Stuart Vorpahl, Jr.,
and I am Christine.
I want to be a trap fisherman like him
when I grow up.
His father and grandfather
were fishermen, too.
They loved the sea,
and they loved to be on their own.
My sister, Susie, and I were born
on Long Island.
We live in the house Daddy lived in
when he was small.
Susie is ten.

I am eleven months older.
My mother, Mary, was also born
and raised on Long Island.
She met my father at a school dance.
They liked each other right away.
I think it was because
both their fathers were fishermen.
Sometimes my mother and Susie
go out with us to the fish traps.
But I am my father's right-hand girl
and the most steady.

We have Bandit, our dog.
He has become a real fishing dog.
He would rather be on the water
than on land.
Then we have Casper and Midnight, our cats,
and Harriet and Ozzie, our crows.
We found Harriet and Ozzie in a potato field.
We were looking for leftover spuds.
There are always some left after
the farmers have gathered the crops.
Harriet and Ozzie were in their nest.
They could not fly yet.
The mother crow had gone to look for food.
An old owl got her.
We saw it happen, so we took
the babies home.

BEWARE OF DOG

We nursed them and trained them.
Daddy has been nursing baby animals
ever since he was a boy.
He even nurses snakes.
Susie and I do the same thing.

We liked the crows best when they were little.
They would cling to our arms.
Now that they are older,
they are curious about all
the new things around them.
They don't want to play as much
or stay around the house.

Next to fishing, I love to draw.

I pack a pad and drawing pen

when we go out on the boat.

I draw things I see along the way.

I like to draw the swallows

sitting on telephone wires.

Once I drew starlings

racing each other across the sky.

But when the catch is good, there is no time

to think about drawing.

I am too busy helping bring in the fish.

Each year around the first of March,

my father starts putting the fish traps

into the water.

Our traps are about seven

miles down from the harbor.

He rows out to his barge.

It is moored not far from shore.

His motorboat, the *Guts*,
and his trap boat are tied to
the barge.
But when he gets ready
to put his traps in,
he unties the boats from the barge.
He anchors them so they won't
drift away. Then he attaches
his friend's big motor to the barge
and goes out to his trap area.
The first thing he does is put
the stakes into the bottom of the bay.
After he has finished setting the traps,
he returns the barge to the harbor.
He again moors the *Guts* and
the trap boat to the barge.
This is where they
stay when they are not in use.

You must have a permit
to fish in trap areas.
We apply to the Army Engineer Corps
every year to get one.
You don't have to pay for the license.
The Army Engineers know
where your traps are.
They send you a nautical chart
with your spot marked on it.
The charts are like maps.
But they use numbers instead of
the names of towns, cities, or states.
Then you must put stakes or poles
into the bay to mark your area.
There must be a sign on one of the stakes
with the owner's name and permit number.
You also must have a light
on the outermost stake
to warn boats that are out at night.

STUART VORPAH JR
AMAGANSETT
U.S. ARMY

Once he has brought everything
he needs to the trap area,
my father starts working the gas pump.
It pumps water
through a hose and pipe.
The pressure of the water
makes a hole in the floor of the bay.
The stake is dropped into the hole.
The sand settles back very quickly
and makes the pole fit snugly.
The next step is to hook
the nets on to the poles.
The trap is now ready to receive the fish.

Trap fishing is a very old way
of catching fish.
We learned how to do it
from the American Indians.

It hasn't changed for hundreds of years.
Fishermen all over the world still
catch fish this way.
As the fish swim along, they enter
the trap through a "leader" net. The
leader is like a long entrance hall.
From there they go into the "kitchen,"
the first room of the trap.

From the kitchen they swim
through a tunnel into the "parlor."
Some fishermen call the sections
of the trap "leader,"
"inner pounds," "funnel," and "box."
The fish become trapped in the parlor
because they can't find their way out.
In the parlor, the fish
group themselves together

into families or schools. Different fish
don't like to mix with one another.
Sometimes, if you leave
the fish in the trap too long,
they get smart and find the way out.
That is why you've got to
get to those nets—rain, wind, or storm.
We are always worrying about bad weather.
All fishermen do.
It might mean losing the fish traps.
When there are no fish to sell, there is
no money to buy food or pay bills.
Once we had a terrible storm
and my father lost two fish traps.
The traps cost a lot of money
and are hard to replace.
We are always mending nets.
They tear easily
with all the traffic underwater.

I know enough now to help mend them.
Daddy builds all his own
boats and equipment.
He made our trap boat longer
so it could hold more fish.
It is made of metal.
I helped him with the welding.
He cut the boat across the middle
and added four feet on each end.

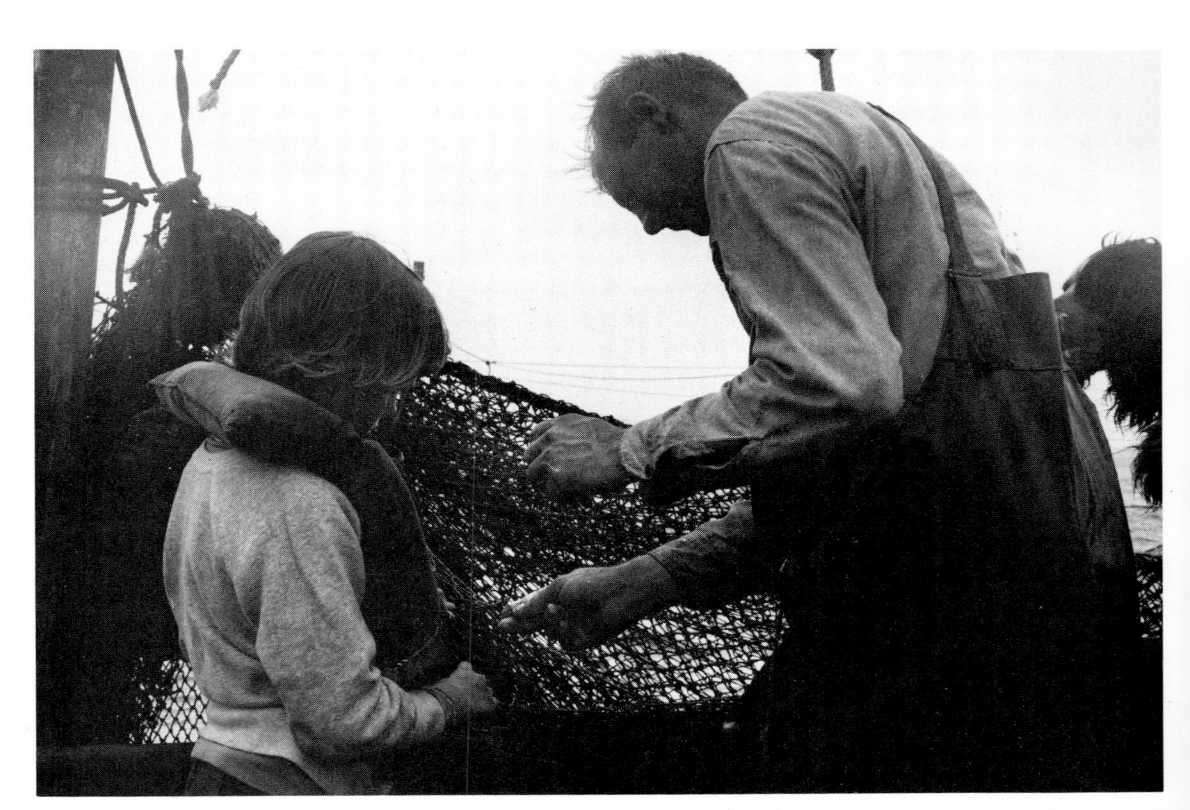

We wore masks when welding.

The brightness of the torch can blind you.

It took five days to do the job.

We had to build a new section.

Then we joined the sections together.

The old pen boards were four feet too short.

I helped Daddy

fasten in the new ones.

The pen boards keep the fish

from sliding all over the boat.

Then we loaded the trap boat onto
the trailer to take it back to the bay.
The metal boat is heavy,
but my father is smart.
He backed the trailer into the water
and slammed on the truck brakes.
The trap boat just rumbled
right off the trailer into the bay.
While he was parking the truck,
I held the trap boat
so it wouldn't drift away with the tide.

Chapter Two

I love the summer.
There is no school.
I can go fishing every day.
My father wakes me around six
in the morning.
I get moving fast.
At that hour the sky is misty and gray.
It's as if the sun isn't sure
about coming out.

The grass is covered with dew.

I eat my breakfast and then fix our lunch.

It is quiet and I am feeling

a little sleepy.

Mother and Susie are still asleep.

My father and I pile the lunch

into the back of the truck,

along with hip boots, mittens, hats,

THIS SHIRT
IS

raincoats, Daddy's oil coat, waders,
scoop nets—and Bandit and me.
And my pad and pencil.
It's very still, except for Bandit's bark.
He gets excited watching us get ready.
He's been fishing for four years,
ever since he was a baby.
The houses seem empty.
There aren't many cars on the road.
There are lots of dogs.
I know some of them by name.
Still, there are always new ones
that I don't know.
Neither of us is saying anything, but inside me,
I wonder if the water will be rough
or if there will be any fish in the traps.
I know that my father
thinks about that all the time.

When there are no fish
or hardly enough to sell,
he is very quiet.
When the nets are full,
he whistles or hums.
It can be very dangerous on the sea
when it is stormy.
One time we were pulling in our catch
and Daddy said
he felt a storm coming.
We hurried to finish up.
Sure enough, we could see
black clouds coming in.
There was a thick black line
to the southwest.
It rose two hundred feet into the sky.
Then little chunks of black clouds
came down like fingers.

Suddenly an explosion of water
blew up into the air.
It was a water spout.
It was like a little tornado.
We tried to steer our boat away,
but we couldn't.
Daddy shouted to me
to get down into the bow
and under the seat.
The water was like a solid tower.
It roared so hard
we couldn't hear anything else.
Then it stopped as fast as it had started.
It got very cold, and the rain came down.
Daddy steered the boat toward land.
I stayed under the seat.
I was afraid to come out
until we reached the shore.

We were just far enough away from the spout.
If we had been any closer,
our boat would have capsized.
Later somebody said that a big yawl
out there had had all its sails torn off.
I guess we were lucky.
Daddy said the sea is like a person.
It has a mind all its own.
Sometimes it does unexpected things.
The day after the storm,
I felt like staying home.
But Daddy said we have to understand
that the sea is also good to us
and that we live by it.
When he told me this,
I changed my mind.
We drove out to the bay.
Our rowboat sat waiting for us

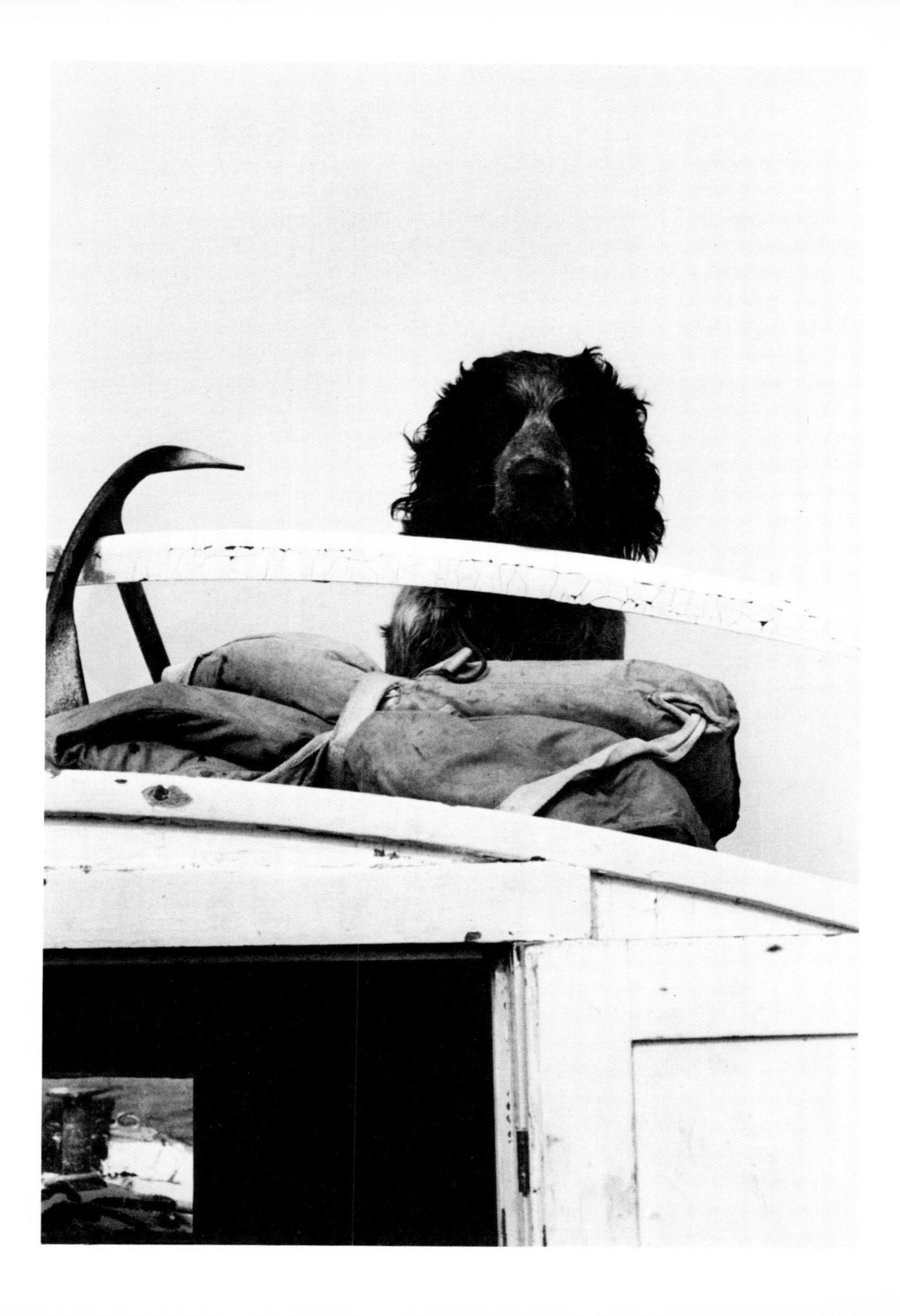

at the edge of the water.
We piled all our things into the rowboat
and rowed out to the barge.
We got into our motorboat, the *Guts,*
and tied our trap boat to it.

Bandit likes to sit on top of the *Guts,*
where he can see all around.
He sits patiently
all the way out to the traps.
He always knows
when we are close to our mark.
Once he got so excited
barking at a large dogfish
that he fell overboard.
A dogfish is a small shark.
There are days when the trap
is filled with them.

My father tried to pull Bandit out
and fell right in after him.
I was scared and didn't know what to do.
My father can't swim well.
Even though he is always on the water,
he doesn't know too much
about being in the water.
Once I dreamed my father was
captain of a large ship.
It was filled with hundreds of people.
They were all fishing.
Suddenly a huge arm with clams
and seaweed growing on it
came out of the ocean.
The arm destroyed the ship,
and everyone drowned except my father.
They were pulling him out
when I woke up hot and sweating.

There are days when the water is
so rough I stay in the *Guts*.
My father gets into the trap boat
alone to lift the traps.
But most days I climb right in
and work along with him.

Chapter Three

On top of each stake there are
pulleys for raising the nets.
There is a rope that is hooked on
to the bottom of the nets.
And there is a winch in the trap boat.
We put the rope through the pulleys.
Then we run the rope to the winch
and wrap the rope around it.
The winch turns around and around
and pulls the rope that is
hooked to the net.
Slowly it lifts up the nets.
As the parlor section with the fish in it
is raised, we steer the trap boat in closer.
We use our scoop nets to get the fish
out of the big net and into the trap boat.

When the net is empty, we let
the rope unwind through the pulleys
until the net drops back into place.
Then we hook up the trap boat
to the *Guts* again
and start back to the harbor.

I have my own scoop net.
My father hollers to me not to put
too many fish into it at once.
It bends and can break easily.
I used to worry about the fish dying,
but my father explained that
we need fish as food.
The fish we catch will feed many people.

One day I scooped up a jellyfish.
It brushed against my upper lip.
My lip stung for a long time.
I felt like crying, but I didn't.
You get used to all these things
out on the water.
And even on the days when we work
very hard, nobody ever gets tired.
The catch is mostly pot luck.
Some days there are schools of fish
swimming right alongside the boat.
There are mackerel, squid, and
butterfish swimming together.
They eat bait, which are little tiny fish.
The fish we can't sell,
like sea robins, mulley kites, bunkers,
baby porgies, and baby butterfish,
we throw back into the sea.

Some days we start gutting the fish
right there in the boat.
Then the sea gulls swarm around.
They eat the fish guts and livers.
Sometimes I find a whole bait inside
when I am gutting fish bellies.
My father really likes this business.
There is nothing else he wants to do.
But in the winter, when everything

is frozen, there is no fishing.
He has to get other jobs.
He tries to work in a shipyard
when he can get the work.
I feel the same way he does about fishing.
It's something about the sea.
When you are away from it,
you can't wait to get back.

When the work is finished
and the catch is good,
my father puts the *Guts*
on automatic pilot
and we breeze through the water.
We eat lunch and sing songs.
This is a sort of reward,
and we both feel good.
We go back to shore.

The trap boat is full of our catch.
We take the fish out of the boat
and pile them into the back of the truck.
Then we cover them with ice.
Sometimes the catch is really special, and
my father lets me ship a box of fish
to the Fulton Fish Market in New York City.
When the check comes in,
it is addressed to me.

THIS SHIRT IS
CYCLABLE

The catch is best in spring and fall.
Before winter comes, the fish
start heading south to North Carolina.
It's like birds migrating.
In the summer months, the fish are feeding
and do not travel long distances.
I love the quiet on the water
with just the sound of the waves
as they touch the sides of the boat.
It's the time I love the best,
alone with a boat full of fish
and Bandit and my father.